Daiana Alves Machado

Food Analysis and Processing Laboratories

Daiana Alves Machado

Food Analysis and Processing Laboratories

Mandatory Curriculum Internship Report Federal Institute of Santa Catarina - câmpus Canoinhas

Imprint

Any brand names and product names mentioned in this book are subject to trademark, brand or patent protection and are trademarks or registered trademarks of their respective holders. The use of brand names, product names, common names, trade names, product descriptions etc. even without a particular marking in this work is in no way to be construed to mean that such names may be regarded as unrestricted in respect of trademark and brand protection legislation and could thus be used by anyone.

Cover image: www.ingimage.com

This book is a translation from the original published under ISBN 978-620-2-80442-4.

Publisher:
Sciencia Scripts
is a trademark of
International Book Market Service Ltd., member of OmniScriptum Publishing Group
17 Meldrum Street, Beau Bassin 71504, Mauritius
Printed at: see last page
ISBN: 978-620-2-83537-4

ACKNOWLEDGEMENTS

To God and to Our Lady of Aparecida, for enlightening and guiding my paths.

To my parents Jair and Dulce and my sister Larissa, for all the love, support and patience at all stages of my life.

To all my family, especially my grandparents João and Lídia, for all the love and prayers that bless me along the way.

To teachers and friends Graciele Viccini Isaka and Luciano Heusser Malfatti, for all the knowledge, support and advice that made all the difference in my academic, professional and personal life.

To Elis C. N. Machado, for his sincere friendship and support at all times.

Josiéli de O. dos S. Veiga, for all the companionship in the management of the IFSC food laboratories - câmpus Canoinhas.

My mentor Mônia Stremel Azevedo, for her trust, guidance, friendship and all the knowledge shared.

To IFSC - câmpus Canoinhas, for the opportunities and support in my professional and academic career.

To my friends, for the encouragement and partnership at all times.

To all who directly or indirectly contributed and supported my achievements.

SUMMARY

1 COMPANY PRESENTATION

The Federal Institute of Education, Science and Technology of Santa Catarina (IFSC) was created as a School of Artistic Apprentices of Santa Catarina by Decree No. 7,566 of September 23, 1909, in Florianopolis. The institution's objective was to provide people from less privileged socioeconomic classes with free and quality professional training (IFSC, 2012).

Currently, besides the Rectory, located in Florianopolis, the IFSC has 22 campuses distributed in 20 cities in the State of Santa Catarina, and distance education centers in Santa Catarina, Rio Grande do Sul, Paraná and São Paulo (IFSC, 2017).

The Canoinhas Campus was inaugurated on February 18, 2011, being the first IFSC teaching unit located in the Northern Santa Catarina Plateau. Operating in the areas of Food Production, Natural Resources, Civil Construction and Information and Communication, it offers concomitant and integrated technical courses to High School, undergraduate and graduate courses, Initial and Continuing Education (FIC), among others (IFSC, 2016).

1.1 Sector

The Food Production axis, which includes the Technical Courses in Food Integrated to High School and the Degree in Food Technology, has six laboratories, which are used for the development of practical classes, for quality control and food processing, necessary for the professional performance of the Food Technologist: laboratories of Physical-Chemical Analysis, Food Analysis, Microbiology, Meat and Vegetable Processing, Bakery and Dairy Processing and Sensory Analysis (under implementation). It also has a warehouse for storage of reagents and shelter for chemical residues (FIGURE 1).

Figure 1 - Sector of development of the activities of the mandatory curricular stage.

Source: The author (2018).

2 INTRODUCTION

The mandatory curricular internship as a Laboratory Technique was developed from January 2017, in the Physical-Chemical Analysis, Food Analysis, Microbiology, Meat and Vegetable Processing, Bakery and Dairy Processing Laboratories and Sensory Analysis (under implementation) of the IFSC - Canoinhas Campus.

This internship aimed to apply the technical knowledge acquired during the graduation in the management of the IFSC food laboratories - Câmpus Canoinhas.

The specific objectives were:

- prepare and follow up the practical classes in the laboratories, as requested by the person in charge;
- Review and improve the Practical Class Application Form;
- elaborate the Chemical Residue Management Plan of the above mentioned laboratories;
- elaborate the Standard Operating Procedure (SOP) of equipment;
- store properly, observing safety standards, the chemical reagents used by laboratories.

In this internship, all the theoretical knowledge acquired in the Food Technology Course was applied in the most different activities carried out, favoring the technical learning and the association of the knowledge with the activities developed in quality control and service rendering laboratories of various sectors of the food area.

3 BIBLIOGRAPHIC REVIEW

The Laboratory Technician, according to the Career Plan of Administrative Technical Positions in Education (PCCTAE), is the professional responsible for:

- prepare chemical solutions and reagents, equipment, glassware and other materials used in experiments;
- proceed with the assembly and organization of experiments gathering all the necessary materials for use in experimental classes and research trials;
- collect samples and data in laboratories or in field activities related to research;
- analyze materials using physical, chemical, physical-chemical and biochemical methods of qualitative and quantitative identification of the components of this material, according to prescribed methodologies;
- clean and conserve the facilities, equipment and materials of the laboratories;
- carry out the acquisition and stock control of consumable and permanent materials of the laboratories;
- take responsibility for the warehouse of the laboratories;
- manage the laboratory with the person responsible for it;
- to use computer programs and software for laboratory control spreadsheets, material acquisition and other institutional processes;
- perform other services of a similar nature and level of complexity integrated to the organizational environment (UNIFESP, 2016).

3.1 Preparation of chemical solutions

The homogeneous mixture between a solute and a solvent, i.e., that presents only one phase, is called a solution. This solution can be liquid, solid or gaseous, and its composition varies according to the need of the experiment or the limitation of variation in composition (RUSSEL, 1994).

Solutions must be prepared carefully and as accurately as possible. For this purpose, chemical reagents with adequate purity index are used, preparing a solution with known molarity, dissolving the reagent in a suitable solvent, which is usually distilled water. To avoid errors in the analytical process, when necessary, the standardization of certain chemical solutions should be performed by titration using appropriate chemical standards in order to know the actual concentration of this solution (RIBEIRO, 2014).

3.2 Preparation of practical classes

The classes are prepared by the teacher with the objective of promoting student learning, so that they can make the most of and relate the content to daily experience (INFORSATO; ROBSON, 2011).

The preparation and organization of practical classes in the laboratory is the responsibility of the Laboratory Technician, who selects all the materials and inputs necessary for the class, as specified by the teacher in a specific form for requesting classes. At the end of the classes, the technician is responsible for checking if the equipment or glassware used has been damaged, storing materials in the appropriate places, as well as maintaining the order and cleanliness of the laboratory for later activities (SOUZA *et al.*, 2014).

3.3 Physico-chemical and microbiological food analysis

Foods are difficult to handle due to their complex organic composition. To determine the components of this constitution, food physicochemical analyses are used, which have significant importance in the evaluation of food quality and safety, in addition to acting as an aid in technological innovations (IAL, 2008).

Instrumental techniques based on physicochemical principles are used in the most diverse food analysis, such as: determination of the centesimal composition;

quantification of components such as vitamins and mineral salts; detection and quantification of chemical contaminants; characterization of food matrices; elaboration of nutritional labeling, among others. These procedures must be performed according to standardized methodologies in order to ensure the reliability of results (IAL, 2008).

Food can also be submitted to microbiological analysis, having basically two research intentions: to detect deteriorating and pathogenic microorganisms present in the product to evaluate their innocuity; and to evaluate the action and efficiency of microorganisms beneficial for food production (FRANCO; LANDGRAF, 2008).

3.4 Food Processing

The food industry aims to extend the period in which the food remains fit for consumption, increase the variety and practicality in the preparation and consumption of these products. The food industries use the principles of food processing in order to achieve these objectives (AGB, 2010).

Food processing, in its most varied applications, provides numerous benefits to the products in terms of availability, *shelf-life* and safety, preventing damage from decomposition and contamination that pose a risk to consumer health, and ensuring their microbiological quality (AGB, 2010).

In addition, the consumer market is increasingly demanding, causing food industries to seek strategies to increase the number of customers and intensify the competitiveness of companies. One of the strategies is the use of technological innovations in the development of new products and processes, ranging from obtaining quality raw materials to the rapid and efficient transport of the final product. The main innovations in the food area occur in the formulation of additives and ingredients, packaging, genetically modified organisms (GMOs), functional and organic food (ABREU, 2012).

3.5 Cleaning and organization of laboratories

In the laboratory environment, experimental studies of any branch of science are performed. By gathering people, equipment, glassware and other materials in the same space, the cleaning of the environment must be performed with special care (COUTO, 2011). The place needs to be clean and free of materials that are not linked to the activities performed in that space (ARAÚJO, 2009).

The daily cleaning of the laboratories is done after the activities performed on site. The cleaning of benches and materials, as well as the organization of waste generated are activities performed by the technician, while the cleaning of the floor and removal of garbage is usually performed by an outsourced employee. The general cleaning is performed in a longer period of time, since in this process the floor, ceiling, wall, window glass and benches of each laboratory are sanitized. The technicians must be present during these procedures, because the cleaning of the benches, equipment and glassware are their responsibility. In addition, the removal or change of place of equipment must be performed carefully by the technician, since he has knowledge of the sensitivity of the equipment (COUTO, 2011).

The outsourced employee should often be guided by the technician in charge of the laboratory regarding the products to be used and the necessary care during cleaning. Normally, the products used for cleaning these spaces include sodium hypochlorite, with concentration between 2 and 5% and ethyl alcohol 70%. (COUTO, 2011).

Regarding laboratory glassware, washing is usually performed with water and neutral detergent, with the help of sponges and brushes specially designed for this function. These materials are rinsed in two stages, the first in running water and the second with distilled water, for complete removal of detergent residues. The drying of the glassware is done in dry places and free of particulate material (dust) or in drying ovens, when these are not volumetric (FELIPE, 2012).

3.6 Chemical Waste Management

The Waste Management Plan (PGR) is a document provided for in ISO 14001, which can be implemented in any type of company, being used in the planning of waste generated in several productive sectors. This document covers several norms and guidelines that guide the best waste management method, from generation to final destination, minimizing environmental impacts and ensuring environmental sustainability. The PGR can address specific standards and guidelines for a class of waste, such as those from chemical sources, through a Chemical Waste Management Plan (PGRQ) (AMANCIO; OBENAUS, 2014).

According to NBR 10004 (ABNT, 1987), chemical waste is solid waste, classified according to potential risks to public health and the environment, and must be properly handled and disposed of.

From the 1990s on, Brazil began to worry about the chemical residues generated in the laboratories of teaching institutions, research centers or service providers, during the research and/or teaching activities developed in these locations. This situation was mainly due to the lack of adequate management of these materials. One of the solutions adopted was to encourage actions aimed at minimizing and treating waste from laboratory activities (MARINHO *et al.*, 2011).

With the increase of environmental problems and impacts resulting from the activities developed in the laboratory, it became indispensable the commitment of the institutions responsible for the management of chemical residues, seeking environmental sustainability. Thus, the need arose for the implementation of a waste management program in these institutions, through the mapping of waste generating procedures and constant training of employees, always in accordance with the standards determined by current legislation (MARINHO *et al.*, 2011).

3.7 Reagent and chemical waste storerooms

NBR 12235/1992 (ABNT, 1992) and NBR 17505-5/2006 (ABNT, 2006) define the basic standards for storage of hazardous solid waste and flammable liquids and fuels in order to protect public health and the environment. According to these standards, the reagents and residues must be stored in an appropriate environment, and the residues are stored temporarily until they are destined to a specialized center for the treatment of hazardous solid residues. These environments must be built in masonry, be tiled and ventilated, with ducts to drain possible waste leaks and dykes to contain these leaks. In addition, the containers where the residues are stored must be properly identified, allowing the rapid identification of the stored compounds.

3.8 Standard Operating Procedures (SOPs)

According to Resolution RDC ANVISA n° 275, of October 21, 2002 (BRAZIL, 2002), Standard Operating Procedure is the "objective written procedure that establishes sequential instructions for routine and specific operations in the production, storage and transport of food".

The importance of the institution of POPs in a research laboratory aims, mainly, to improve the preparation and harmonization of research processes and methods; to improve the quality and efficiency in training of researchers; to promote the professionalism and credibility of the institution; to ensure the quality and repeatability of laboratory results, among others (BARBOSA *et al.*, 2011).

The SOP should describe in detail the procedure in question, ensuring uniformity in production or service provision. The document must contain some basic items, such as header, code, title, logo of the institution or company, responsible persons, dates of preparation and/or revision, description of procedures, among other pertinent information. These documents must be accessible to the laboratory users, either in printed or electronic form, and must always be updated (BARBOSA *et al.*, 2011).

3.9 Bidding process

"Bidding is a formal administrative procedure in which the Public Administration calls, by means of conditions established in a proper act [...], companies interested in the presentation of proposals for the offering of goods and services". (BRAZIL, 2010, p. 19).

Bids and administrative contracts related to services, works, purchases and leases at the national, state, district and municipal levels are regulated by Law No. 8,666 of June 21, 1993. It determines that, in the Public Administration, contracts made with third parties must be preceded by a bidding process (BARRETO, 2008).

The objective of the bidding process is to select the most advantageous contract proposal, ensuring compliance with the constitutional principle of isonomy, so that all interested parties have the same opportunity, allowing a greater number of participants in the contest (BRAZIL, 2010).The bidding process is divided into six modalities: competition, price taking, invitation, bidding, auction and trading. The first five modalities are generally defined according to the value of the future contract, with the exception of the auction which is defined according to the good or service to be bid (BARRETO, 2008).In all bidding modalities there are two phases, the internal and the external phase. The internal phase includes the definition of the item specification, the estimate of prices, indication of resources, choice of modality and type of bidding process and preparation of the bidding notice according to specific standards. In the external phase, the stages of publication of the bid invitation or letter, qualification of the bidding companies, analysis and classification of the price proposals according to the criteria of the bid invitation, judgment, adjudication, homologation, contracting and execution of the object are included (BARRETO, 2008).

3.10 Inventories

Inventory is a survey and identification of materials to prove their physical existence, as a way to control and preserve the public patrimony, proving the total balance at the end of the fiscal year (PARISIO, 2014).During the physical inventory, the accounting and administrative records and controls are updated; the assets of the collection are proven in terms of species, value and quantity; the conditions of conservation of permanent materials in use and the needs for repairs, maintenance or casualties are identified; and the need for the asset in that sector or unit is assessed (KLIPPEL *et al.*, 2013).The inventory can be done in a general or rotating way. The general modality is carried out at the end of each organ's exercise, including, at once, all the registered items. In the case of a rotating inventory, the asset conferences are held each month with a certain group of materials, avoiding the paralysis of the sectors and facilitating the identification and correction of problems in a timely manner (PARISIO, 2014).

3.11 Stock control

According to Davis *et al.* (2001, apud CUSIN *et al.*, 2012, p. 3), the definition of stock can be interpreted as the quantification of any materials or goods used in a public or private institution. The ideal in an organization would be not to have unnecessary inventory, but this is not always possible, since inventory is a way to guarantee a safety margin for the constant functioning of the company.Stock control is a process of registration, inspection and management of the flow of incoming and outgoing materials and products of the company. Stock planning is the most efficient way to avoid wasting financial resources and products, ensuring the supply of demand without affecting the budget (SEBRAE, [20--?]).

4 ACTIVITIES DEVELOPED DURING THE INTERNSHIP

The activities of the mandatory curricular internship as a Laboratory Technique were developed from January 2017, in the Physical-Chemical Analysis, Food Analysis, Microbiology, Meat and Vegetable Processing, Bakery and Dairy Processing Laboratories and Sensory Analysis (under implementation) of the IFSC - Canoinhas Campus.

All the procedures and activities developed were supervised by the Immediate Leader responsible, and the knowledge acquired in the Food Technology Superior Course was of fundamental importance for the theoretical basis and satisfactory development of all the activities.

The main activities developed are described below, with emphasis on the preparation of experiments for food analysis and processing, which helped to understand the procedures commonly employed in food service and quality control laboratories.

4.1 Preparation and follow-up of practical classes

The laboratories are used by almost all the technological axes of the Canoinhas lamp, with the exception of the computer area, with practical classes that go from basic chemistry to food production.

The teachers make the practical class request through a specific form (APPENDIX A), where the materials, reagents, equipment, number of teams, the laboratory and the time when the class will be held, among other information. In addition, there is a place destined exclusively for the identification of the residues generated during the experiments, besides the possibility of suggestion, by the responsible teacher, about the possible forms of treatment of this residue. This form is normally delivered two days before the practical class.

After receiving the request, the class is prepared in advance, where kits are organized for each team (FIGURE 2), meeting all the needs of the teacher. If necessary, the methodology is tested before the class, especially when it is a new procedure, in order to guarantee the efficiency of the experiments.

Figure 2 - Example of practical classroom kits.

Source: The author (2018).

During some classes, mainly of the integrated course classes, the monitoring of the classes was done, helping the teachers in the experiments and in the fulfillment of the safety norms.

After the end of the practice, the space used was organized, performing the washing of glassware, cleaning of benches and equipment, in addition to the proper disposal of residues, chemical or not, generated during the experiment. In case any laboratory occurrence was observed, such as broken glassware or damaged equipment, it was registered in a spreadsheet of occurrences for future control.

The experimental classes that were organized in this period can be divided into two groups, the first group consisting of food analysis classes and the second of food processing classes.

4.1.1 Practical food analysis classes

Food analyses include physical-chemical and microbiological analyses, whether for product characterization, determination and quantification of food matrix components or product quality evaluation, according to the specific purpose.

The experimental classes were prepared, organized and tested at the request of the person responsible for the experiment. Usually, the request is accompanied by a script where all the procedures are described, or with the identification of the methodology to be followed, helping in the preparation of the class.

It is worth mentioning that all the practical classes are prepared in accordance with the safety standards of the laboratories, making use of collective protection equipment (EPC) such as exhaustion chapel, and personal protection equipment (PPE), such as dust cover or coat, gloves, mask and goggles.

4.1.1 Physico-chemical food analysis

The physical-chemical analyses have a weekly flow of realization, and the classes usually take place at least twice a week, with different classes and procedures.

Among the analyses performed, the main ones are those of determination of:
- residual chlorine in water;
- oxygen dissolved in water;
- humidity by drying in an oven at 105 °C;
- total acidity in milks, vegetable oils, fruits;
- waste by incineration or ashes;
- chlorides by volumetry;
- lipids or ether extract by direct extraction in Goldfish;
- proteins by the MicroKjeldahl Method;
- citric acid in fruits;

- foreign bodies in honeys;

- hardness of carbonates and non-carbonates in water;

- Total alkalinity in water by volume;

- pH and density in several products;

- Refractometry soluble solids;

- peroxide and saponification index in oils;

- vitamin C in juices;

- milk fat by the Gerber Method;

- Centesimal composition of milks by EcoMilk Total;

- somatic cells in milk by EcoMilk Scan;

- starch in milks, jellies, among others.

Figure 3 presents a kit for the determination of pH in different solutions (a), determination of total acidity in milks (b), determination of foreign bodies in honeys (c) and visualization of foreign body observed in the stereomicroscope (d).

Figure 3 - Food physicochemical analysis performed in practical classes.

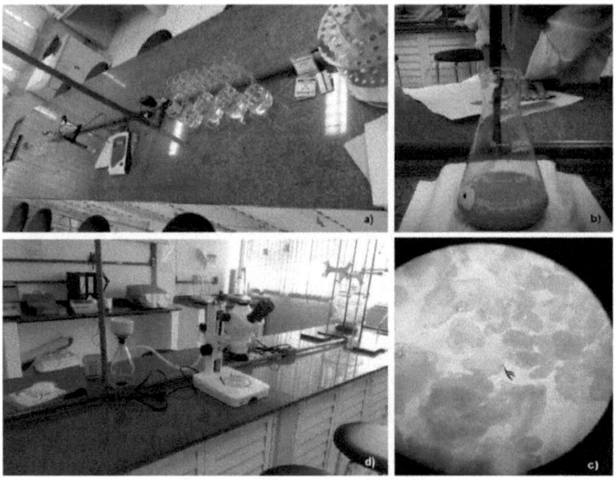

Source: The author (2018).

The performance of a physical-chemical analysis usually depends on chemical reagent solutions or the early preparation of some materials and even the availability of some food matrices. Regarding the above mentioned analyses, some chemical reagents solutions are necessary, with different molar concentrations (M or mol/L), such as sodium hydroxide, sulfuric acid, hydrochloric acid, ethylenediaminetetraacetic acid (EDTA), potassium hydroxide, iodine, silver nitrate, potassium permanganate, sodium thiosulphate, among others. All the reagents are prepared in a gas exhaustion chapel, based on the basic knowledge of analytical chemistry acquired during the academic training.

In some analyses, especially in procedures involving titration, it is necessary to use an indicator to facilitate the visualization of the reaction turning point. The main indicators used in the Food Analysis and Physical-Chemical Analysis Laboratories are phenolphthalein, methyl orange, starch solution and methyl red.

The indicators and chemical solutions are prepared preferably on the day of the experiment, or one day in advance, from the teacher's request, since the reagents are stored in a specific warehouse. The reagents are taken to the laboratory, where the solutions are prepared as requested, following all the care to ensure the quality of the reagents and the personal safety of the technician. All movements of reagents inside the warehouse are recorded in a control sheet.

In addition, some classes use equipment that requires prior calibration, such as pHmeter, or early programming of commands, such as temperature regulation in greenhouses and heating bath with circulation, spectrophotometers, among others.

4.1.1.2 Microbiological food analysis

Microbiological analyses are performed at the Microbiology Laboratory. The frequency of use of the laboratory depends mainly on the curricular matrixes of the courses, since it is generally used with greater flow by the General Microbiology and Food Microbiology Curricular Units (UC). It is also attended during product development in some CUs such as New Product Development and Integration Project, and in the execution of research projects.

Among the main analyses performed in the laboratory are those of:

- Quantification of Mesophilic Aerobic Bacteria;
- quantification of moulds and yeasts;
- quantification of Total Coliforms and Thermotolerants;
- Positive Coagulase Staphylococci quantification;
- quantification of *Bacillus cereus*;
- research of *Salmonella* spp;
- evaluation of fungi control by different chemical agents;
- research of endophytic fungi in grains;
- evaluation of microbial load on different surfaces and environments;
- Price experiment;
- simple and differential coloration (Gram);
- sowing, pricking out and cultivation of bacteria and fungi.

Figure 4 presents the results of microbiological analysis of endophytic fungi in grains (a), evaluation of fungi control by different chemical agents (b), quantification of thermotolerant coliforms (c), quantification of total coliforms (d), research of *Salmonella* spp. (e) and quantification of coagulase positive staphylococci (f).

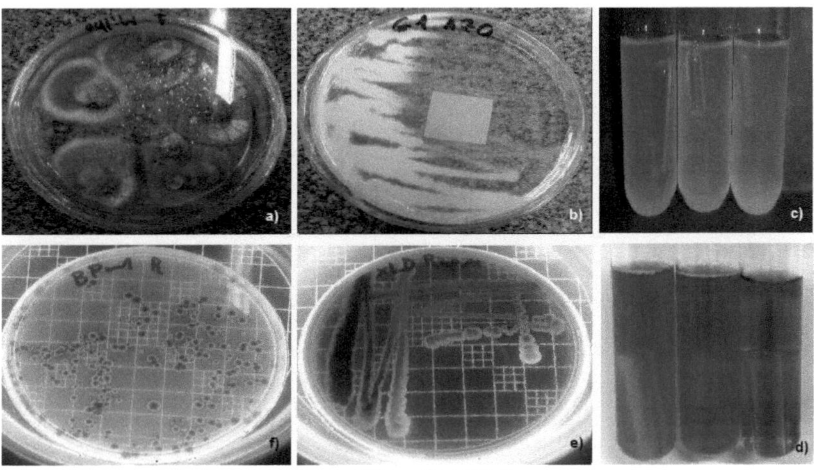

Source: The author (2018).

Microbiological analyses are totally dependent on the anticipated preparation of materials, because to perform them it is necessary to use culture media and sterile materials.

The culture media were prepared according to the manufacturer's guidelines, following the recommendations for adding supplements such as egg yolk, polymyxin, potassium tellurite, iodine solution; and for autoclaving or not the culture media. If the culture medium cannot be autoclaved, it is usually prepared in a heating bath until the medium is completely dissolved. The storage is refrigerated or in dark environment until the use in the experiment.

Some of the glassware and materials such as platinum needles, tweezers, scalpel, spatulas, must be autoclaved previously for use in the experiments. These materials are previously prepared and placed in the kits for each team, accompanied by the other necessary materials, as requested by the teacher. In the same way, some equipment must be programmed before the practical class, such as incubation greenhouses.

After the classes, all materials used are autoclaved at 121 ºC for 20 to 30 minutes in order to guarantee the complete elimination of microbiota from the materials and safety of users.

4.1.2 Practical food processing classes

The practical food processing classes are held at the Bakery and Dairy Laboratories and Meat and Vegetable Laboratories, by students from various courses offered on the campus, as well as projects such as SIM Women, FIC courses and workshops offered by the internal or external public.

Among the main practical classes held in these laboratories are:

- elaboration of minas frescal cheese;
- dairy beverage elaboration;
- elaboration of ice creams;
- elaboration of jellies and cut candies;
- elaboration of minimally processed vegetables;
- elaboration of banana-passa and dry tomato;
- elaboration of preserved vegetables;
- elaboration of breads, pizzas and cookies;
- elaboration of fruit juices;
- elaboration of meat sausages;
- elaboration of matured meats;
- elaboration of milk jam;
- elaboration of sauerkraut;
- craft beer elaboration;
- quantification of gluten content in flours;
- determination of water retention capacity in meat.

Figure 5 presents the products processed in the practical classes of jelly (a), French breads (b), *cookies* (c), ice cream (d), minimally processed vegetables (e) and mature meat (f).

Figure 5 - Products elaborated in the practical classes of food processing.

Source: The author (2018).

All classes were prepared from the class request, with the organization of kits on the benches. In addition, the equipments such as bread growing oven, drying oven with forced air circulation, turbo electric oven, are programmed according to the orientation of the class applicant. If necessary, support is provided during the execution of the laboratory activities.

The practical activities simulate the technologies applied in industrial processes, relating theory and practice. During the preparation and organization of classes, one can learn and relate conditions of anticipated preparation that are not always observed during the execution of the class. In some cases, this knowledge is expanded when there is follow-up during methodology tests.

4.2 Cleaning and organization of laboratories

The laboratories were cleaned whenever there was some kind of activity in the environment, or every two days, when little used. This cleaning was done with sodium hypochlorite from 2 to 5% and ethylic alcohol 70% over the benches and equipment, besides the cleaning of the floors that was done by an outsourced employee.

At least twice a year a general cleaning was carried out, with the help of outsourced employees. In this process, the cleaning of the wall, ceiling, floor, equipment, cabinets, etc. takes place, aiming at the complete hygienization of the environment.

After cleaning, an efficiency test is performed, i.e., Petri dishes with culture medium are placed for room air sedimentation, which are then incubated and quantified according to standard methodologies.

In addition, the food that is stored in the cabinets is removed, cleaned and stored again, with the exception of expired products and those that are not properly identified.

In the case of cleaning the laboratory glassware, washing is performed with neutral detergent, using sponges and brushes. After cleaning, they are rinsed in running water until there are no detergent residues. Afterwards, they are rinsed with distilled water two to three times. After the glassware are dried in greenhouses or in a dry environment and free of particulate material.

4.3 Chemical Waste Management

The chemical residues generated in activities developed in the laboratories are disposed of or stored according to the guidelines determined in literature (FONSECA, 2009; LASSALI, 2003; SALES E SILVA, 2003; TOMAZINI; 2006), current legislation (BRAZIL, 1981; BRAZIL, 2004; BRAZIL, 2010) and Chemical Product Safety Data Sheets (MSDS) provided by companies marketing chemical reagents.

Some of the compounds formed in class (hydrochloric acid solutions, sodium hydroxide, potassium hydroxide, potassium permanganate (after precipitation of manganese dioxide)) can only be neutralized and discarded in the sink under running water. Lassali (2003) states that hazardous and non-hazardous waste that can be neutralized and destroyed in the generating laboratory, should not be accumulated, since the treatment of small amounts of waste is easier and offers less risk to the person responsible for this process.

Some wastes are more complex and need more elaborate treatment. Normally, these are stored in a high density, high molecular weight polyethylene container of fifty liters, separated in six classes (mixed solutions of basic character, mixed solutions of acid character, halogenated organic solvents, non halogenated organic solvents, organic solvent solutions and solutions of salts of transition metals), according to the chemical compatibility of these components (BRAZIL, 2004). These containers are located in the Physical-Chemical Analysis Laboratory (FIGURE 6), making handling easier, since these classes of solutions are frequently generated in laboratory activities.

Figure 6 - Plastic containers for storage of chemical residues.

Source: The author (2018).

However, there are chemical residues that do not fit these classifications, due to peculiarities and chemical incompatibility. In these cases, the residues are stored in a five liter high density polyethylene container, respecting the chemical compatibility of the components, and stored in a warehouse for chemical residues. These containers remain stored in the warehouse until a significantly large amount of waste is reached to be forwarded to a company specialized in waste treatment.

In view of the significant increase in chemical residues and the absence of space to store them, we have started the elaboration of a Chemical Residue Management Plan. The plan is still in the initial phase of elaboration, but all the necessary bibliographic survey about the subject has already been done. After the elaboration of the document, it will be evaluated by a professional of the chemical area, for evaluation and contributions, to then start the implantation phase of the residues management in the IFSC laboratories - câmpus Canoinhas.

4.4 Reagent Warehouse

Previously, the chemical reagents were stored in laboratory cabinets, not being performed a proper separation and offering risks to users. After the elaboration of the technical report of the IFSC - Canoinhas laboratories, it was recommended the reallocation of these reagents in an adequate place, that is, in an environment built in masonry, tiled and ventilated, with drains to drain possible residues and a reservoir with masonry walls to contain these leaks (ABNT, 1992, 2006).

Thus, after dialogue with the managers of the lamp, a bathroom that was unused was reformed, becoming the warehouse of reagents, seeking to meet all the standards established by NBR 12235/1992 (ABNT, 1992) and NBR 17505-5/2006 (ABNT, 2006). The electrical installations were removed from the environment to avoid possible short circuits in the network, since this warehouse would have several flammable compounds. Besides, wooden shelves were built, painted with oil paint, resistant and fixed to the walls.

While all the adjustments were made in the reagent warehouse, all the reagents stored in the laboratory were collected, organized in an electronic spreadsheet. Then, these chemical reagents were separated into seven classes according to the chemical incompatibility of these substances: acids, bases, salts, metals, solvents, toxic and low reactivity, according to the guidelines described in the MSDS, UNIFAL-MG (2017) and UNIFESP (2017).

After classification, the reagents were renumbered according to the updated reagent list and transported to the reagent room, which were stored according to the distribution made after the reagent incompatibility study.

After storage, the shelves were identified, with the name of the defined class and all the reagent codes belonging to that shelf. A control sheet of reagents entrance and exit of the warehouse was elaborated, where the name of the reagent, the code, the quantity, the laboratory where it will be taken, and the responsible for the withdrawal and return of the flasks are registered. This control sheet is in a folder with the updated list of reagents, located at the entrance of the warehouse, seeking to favor filling.

4.5 Standard Operating Procedures (SOPs)

All the equipment of a laboratory must have a specific POP, in order to facilitate the use of the equipment, besides standardizing the processes and procedures. During the internship, a survey was made of the equipment that did not yet have POP, which was found to be the majority of the equipment belonging to the six laboratories.

From this, all pertinent information in the instruction manuals provided by the manufacturers during the purchase of these products was sought, as well as additional details with the manufacturers and users of these equipment. From this information, POPs containing the necessary detailed information were elaborated, to later fix these documents near the equipment to be used.

Appendix B presents a POP model used for the BEL brand analytical scales.

In this activity, it could be observed that some implementations, in the future, may be necessary in the elaborated POPs, because some of these products had not yet been used, and the efficiency of a POP is commonly verified during its use.

4.6 Bidding process

The purchase of some materials and equipment for the laboratories is made by the Laboratory Technician through a bidding process (UNIFESP, 2016). This process is basically divided into three stages: estimation, proposal acceptance and material commitment (BRAZIL, 2010).

Estimates are sent via e-mail to include or not materials in the process. If you wish to include some material, three estimates should be sent from different companies and that do not vary 35% of the value between one and the other, besides the general description of the product, the detailed specification and the quantity to be requested.

After estimates, it was observed that the items are included in the corresponding trading session and are open for the bids of interested companies. At the end of this stage, there is a winning company, with the exception of the process considered desert, due to the absence of a participating company. The nailmaker responsible for the trading floor will send the company's proposal for technical evaluation of the applicant. Then a detailed evaluation of the proposal sent is carried out and a decision is made to accept or not the item offered by the company. After the proposal is accepted, the trading floor follows for ratification and publication on the Integrated System of Patrimony, Administration and Contracts (SIPAC).

The last step of the process is the commitment of the materials of interest belonging to the trading floor in force in the Price Registration System (SRP) of SIPAC. The commitment is made through a specific form available on the IFSC - Canoinhas website, where all the information pertinent to the trading session to which the item of interest belongs and the origin of the resource for acquiring the material, as well as the purchase justification are provided.

After all these steps, the process is finished, the materials are acquired and delivered to the lamp, and the conference is held to investigate possible inconsistencies with what was offered by the company.

4.7 Inventories

All IFSC permanent materials have a number of assets that is checked annually in the inventory process, which is formed by an Inventory Commission made up of persons in charge of various sectors and who carry out the survey of the equity items registered in the lamp.

In the laboratories, the inventory is performed by the laboratory techniques that have technical knowledge of the registered items, optimizing the conference process. If any inconsistency is observed in the spreadsheets extracted from the system, the Heritage Sector is communicated so that corrections or other appropriate measures can be made.

In addition to the IFSC asset inventory, the internal inventory of the laboratories was also carried out, usually twice a year, during academic recesses, where all materials and equipment belonging to the laboratory, whether permanent or consumption, are accounted for. In the year 2017, the IFSC Food Production Area Laboratories assets inventory was carried out four times, due to the general audit of the IFSC assets and regularization of the internal assets of Canoinhas.

Klippel *et al.* (2013) consider that this process is of extreme importance in the public sector, since some inconsistencies in the asset listing can be identified. During the asset inventory survey, it was observed that some items were allocated in locations different from those registered in the system, which allowed their location and regularization. In addition, some observations were added regarding the conservation conditions of some permanent materials in use, equipment that are separated for maintenance or that can be included in the process of viability analysis of asset retirement.

4.8 Stock control

The stock of the laboratories is controlled through electronic and manual spreadsheets, where all the consumption items that enter or leave the laboratories are recorded. From these records it is possible to plan which materials need to be estimated on a trading floor and predict how much money will be needed to supply the basic items needed in the laboratories.

The stock control has total influence on the functioning of the laboratories, because in case there is any failure in the registration of materials, you can stop purchasing some items that are running out or buy unnecessary materials.

5 FINAL CONSIDERATIONS

All the activities developed throughout the internship were carried out with commitment and dedication, however some difficulties were encountered during the course of the internship and may hinder the management of a laboratory and even a company, but especially when it comes to public service.

Some public sector activities are very bureaucratic and end up hindering the processes. A practical example is the bidding processes, which do not always make it possible to include some materials in the trading sessions due to the difficulty of obtaining budgets that meet the necessary requirements. The food sector is very dependent on the companies in this regard, especially with regard to the most expensive and reasonably sophisticated equipment.

Another factor that negatively influences the management of the laboratories is the difficulty in obtaining financial resources for the implementation of improvements and even the simple maintenance of the laboratories. The constant cuts in investments end up impacting not only on the development of the laboratory's activities, but also on the difficulty of executing research projects carried out on the lamp.

The Waste Management Plan is still in its initial phase, and the main difficulty was to reconcile the demand for practical classes and all other activities developed, in addition to the fact that this year we underwent internal audit and adjustments of laboratories that made it difficult to meet the initial schedule for the implementation of the PGR.

Despite the difficulties and delay, we have improved the Application Form for Practical Classes that will help in the process of managing chemical waste, through the correct identification and disposal of waste generated. It is hoped that this improvement in the form will contribute to the reduction of the volume of waste generated in classes.

It was also possible to perform all the readjustment of reagents and chemical residues in appropriate places, minimizing the risks to the users of the laboratories. In addition, the new arrangement of the reagents allows better location and greater control of the use of these substances.

The activities carried out throughout the development of the internship made possible the application of the knowledge acquired in theoretical classes during the Food Technology Course, such as chemical reactions, quality control in the food industry, food processing techniques, in the daily life of a Laboratory Technician. In addition, it was possible to understand the complexity and responsibility of a quality control laboratory in the food industry, because all the procedures, pre- and post-analysis, must be carried out with great attention and commitment, in addition to all the records and controls that must be made and documented in order to ensure the efficiency and reliability of the analyses performed.

In general, both the achievements and the difficulties encountered have contributed to my academic and professional formation, as they have made the theoretical contents have practical application, besides providing a greater understanding that every detail, regardless of the activity that is being developed, must be followed according to certain norms and attention, care and curiosity must always be taken in all areas of activity, always seeking constant improvement.

REFERENCES

ABREU, A. The importance of technological innovation in the food industry: a case study in a large company. In: National Meeting of Production Engineering, 32., 2012, Bento Gonçalves. **Annals...** Bento Gonçalves: ENEGEP, 2012.

AGB. **Principles of unitary operations in food processing.** Federal University of Pará, Pará, 2010. Available at: ppgcta.propesp.ufpa.br/ARCHIVES/documents/Operations%20unit%20no%20proces samento%20de%20alimentos.pdf. Access on: 10 jan. 2018.

BRAZILIAN ASSOCIATION OF TECHNICAL STANDARDS. **NBR 10.004**: Solid Waste: classification. Rio de Janeiro, 1987. 63 p.

BRAZILIAN ASSOCIATION OF TECHNICAL STANDARDS. **NBR 12235**: Storage of hazardous solid waste. Rio de Janeiro, 1992. 14 p.

BRAZILIAN ASSOCIATION OF TECHNICAL STANDARDS. **NBR 17505-5**: Storage of flammable liquids and fuels. Rio de Janeiro, 2006. 25 p.

AMANCE, J. M.; OBENAUS, L. C. Chemical Waste Management Plan (PGRQ). In: IFSC Research, Extension and Innovation Seminar, 4., 2014, Gaspar. **Annals...** Gaspar: SEPEI, 2014.

ARAÚJO, S. A. **Good Practices in the Basic Area of Biological Sciences and Health Laboratory.** Potiguar University. Natal, 2009.

BARBOSA, C. M. *et al.* The importance of standard operating procedures (SOPs) for clinical research centers. **Rev. Assoc. Med. Bras.** , 2011, v. 57, n. 2, p. 134-135.

BARRETO, M. L. T. **Bids:** basic notions. Universidade Federal do Pampa. 2008.

BRAZIL. Law No. 6,938, of August 31, 1981. Provides on the National Environmental Policy, its purposes and mechanisms of formulation and application, and makes other provisions. **Federal Official Gazette**: Brasília, DF, Section 1, n. 167, p. 16509, 02 Sep. 1981.

BRAZIL. Law No. 12,305, of August 2, 2010. Establishes the National Policy on Solid Waste; amends Law No. 9605 of February 12, 1998; and makes provisions. **Official Gazette**: Brasília, DF, Section 1, n. 147, p. 3, 03 Aug. 2010.

BRAZIL. Ministry of Health. Resolution RDC No. 275, October 21, 2002. Provides on the Technical Regulation of Standard Operating Procedures applied to Food Producing/Industrializing Establishments and the Checklist of Good Manufacturing Practices in Food Producing/Industrializing Establishments. **Federal Official Gazette**: Brasília, DF, Section 1, n. 206, p. 126, 23 Oct. 2002.

BRAZIL. Ministry of Health. Resolution RDC No. 306, December 7, 2004. Provides on the Technical Regulation for the management of health services waste. **Official Journal of the Union**: Brasília, DF, Section 1, n. 237, p. 49, 10 Dec. 2004.

BRAZIL. Court of Auditors of the Union. Tenders **and contracts:** guidelines and jurisprudence of the TCU/Court of Auditors of the Union. 4 ed. Brasília: TCU, General Secretariat of the Presidency: Federal Senate, Special Secretariat of Publishing and Publications, 2010.

COUTO, H. A. R. **Cleaning in laboratories:** procedures and special care. Manaus: Embrapa Western Amazon, 2011.

CUSIN, R.; SILVA, V. V. M.; NEUMANN, C. S. R. Control and management of the consumption of materials in stock in the warehouse of a quality control laboratory. In: National Meeting of Production Engineering, 32., 2012, Bento Gonçalves. **Annals...** Bento Gonçalves: ENEGEP, 2012.

DAVIS, M. M.; AQUILANO, N. J.; CHASE, R. B. **Fundamentals of Production Management.** Porto Alegre: Bookman, 2001.

FELIPE, C. **Curricular internship report**: supervised internship. Federal Technological University of Paraná. Campo Mourão, 2012.

FONSECA, J. C. L. **Manual for hazardous waste management.** São Paulo: Academic Culture, 2009.

FRANCO, B. D. G. de M.; LANDGRAF, M. **Microbiologia dos alimentos.** São Paulo: Atheneu Publishing, 2008. 182 p.

INFORSATO, E. C.; ROBSON, A. S. The preparation of classes. In: UNIVERSIDADE ESTADUAL PAULISTA. Prograd. **Caderno de Formação:** formação de professores didática geral. São Paulo: Cultura Acadêmica, 2011, p. 86-99, v. 9.

ADOLFO LUTZ INSTITUTE (São Paulo). Physical-chemical methods for food analysis. Coordinators Odair Zenebon, Neus Sadocco Pascuet and Paulo Tiglea - São Paulo: Instituto Adolfo Lutz, 2008. p. 1020.

FEDERAL INSTITUTE OF SANTA CATARINA. **History.** IFSC Portal, 2012. Available at: http://www.ifsc.edu.br/historico. Access on: 10 Aug. 2017.

FEDERAL INSTITUTE OF SANTA CATARINA. **Câmpus/Polos.** IFSC Portal, 2017. Available at: http://www.ifsc.edu.br/campi. Access on: 10 Aug. 2017.

FEDERAL INSTITUTE OF SANTA CATARINA. **IFSC celebrates five years of Câmpus Canoinhas.** IFSC Portal, 2016. Available at: http://www.ifsc.edu.br/campus-canoinhas/5659-ifsc-comemora-cinco-anos-do-campus-canoinhas. Access on: 10 Aug. 2017.

KLIPPEL, A. L. G. S.; CRUZ, G. M. R.; SILVA, A. M. **Report of the physical inventory of assets of the Cassiano Antônio Moraes University Hospital - HUCAN.** Federal University of Espírito Santo. Viçosa, 2013.

LASSALI, T. A. F. **Chemical waste management:** general rules and procedures. University of São Paulo, Chemical Waste Laboratory, Ribeirão Preto, 2003.

MARINHO, C. C. *et al.* Chemical Waste Management in a Teaching and Research Laboratory: the Experience of the Limnology Laboratory of UFRJ. **Ecl. Química,** São Paulo, v. 36, n. 2, p. 85-104, 2011.

PARISIO, T. **Management of materials and assets in public service.** School of Accounts of the Court of Municipalities. Goiás, 2014.

RIBEIRO, B. R. **Supervised internship report:** quality control - Quimitol. Federal Technological University of Paraná. Toledo, 2014.

RUSSEL, J. B. **General Chemistry.** Volume 1. 2nd edition, São Paulo: Makron Books, 1994.

SALES, J. E. Y.; SILVA, E. B. **IBILCE-UNESP Guide for neutralization and disposal of hazardous chemical waste.** Universidade Estadual de São Paulo, São José do Rio Preto, 2003. Available at: http://www.qca.ibilce.unesp.br/prevencao/protocolo.htm. Access on: 10 Aug. 2017.

SEBRAE. **The importance of stock control.** [20--?]. Available at: http://www.sebrae.com.br. Access in: 05 Oct. 2017.

SOUZA, L. C. M. *et al.* Management in laboratories and methodologies of practical classes for the discipline of General Chemistry applied to Engineering Courses. In: Simpósio Nacional de Ensino de Ciência e Tecnologia, 4., 2014, Ponta Grossa. **Annals...** Ponta Grossa: SINECT, 2014.

TOMAZINI, F. M. **Waste disposal guidance booklet in FMUSP-HC System.** University of São Paulo, Medical School of the Federal University of São Paulo, 2006. Available at: www2.fm.usp.br/gdc/docs/cep_5_grss_2_cartilha.pdf. Accessed on: 10 Aug. 2017.

FEDERAL UNIVERSITY OF ALFENAS-MG. **Chemical Incompatibility.** Permanent Commission for the Prevention and Control of Environmental Risks, Alfenas, MG, 2017. Available at: http://www.unifal-mg.edu.br/riscosambientais/incompatibilidadequimica. Access in: 01 nov. 2017.

UNIVERSIDADE FEDERAL DE SÃO PAULO. **List of incompatible substances.** 2017. Available at:

https://www.unifesp.br/campus/san7/images/pdfs/Tabela_Incompatibilidade.pdf.
Access on: 27 Dec. 2017.

UNIVERSIDADE FEDERAL DE SÃO PAULO. **Career Plan for Technical-administrative Positions in Education:** description of Level D positions. Developed by the Internal Commission of the Career Plan for Technical-administrative Positions in Education, 2013-2016. Presents the description of the positions of Administrative-Technician in Education at Level D. Available at:
http://www2.unifesp.br/reitoria/orgaos/comissoes/cis/descricao-dos-cargos-do-pcctae/nivel-d/view. Access on: Dec 27, 2017.

APPENDIX A - Application Form for Practical Classes

INSTITUTO FEDERAL DE EDUCAÇÃO, CIÊNCIA E TECNOLOGIA DE SANTA CATARINA
CAMPUS CANOINHAS

Solicitação de Aula Prática

*(Este formulário deverá ser entregue devidamente preenchido e em mãos as
Técnicas de Laboratório, pelo menos **03 dias úteis** antes da atividade prática)*

Docente:			
Disciplina/Curso:			
Título da Prática:			
Data da aula:	Horário:	Laboratório:	Número de equipes:

Observações: ____

____

Quantidade do reagente	Reagentes / soluções (concentração – evitar fórmulas)	Quantidade por equipe	Equipamentos / materiais / vidrarias / EPI´s

ATENÇÃO:
- Considerar na aula experimental o tempo necessário para que o aluno possa lavar/descartar o material por ele utilizado.
- Caso haja na aula material, reagente ou resíduo a ser armazenado, identifique-o corretamente com descrição, nome do responsável e data.

RESÍDUO – COMPOSIÇÃO PROVÁVEL	*D/S	QUANTIDADE	LOCAL

* D = descartado; S = segregado.

INFORMAÇÕES ADICIONAIS E SUGESTÕES PARA TRATAMENTO DOS RESÍDUOS

____
Assinatura Docente

Para uso exclusivo da(s) Técnica(s) de Laboratório:

Data de recebimento da solicitação: ___/___/____

Observações:____

____

Data: ___/___/____   _______________________________
                                        Assinatura Técnica

APPENDIX B - Standard Operating Procedure of the BEL Analytical Scale

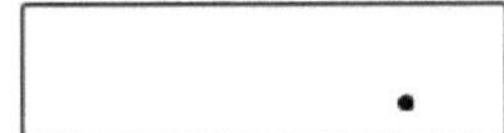	PROCEDIMENTO OPERACIONAL PADRÃO	POP Nº 017
SC Canoinhas		

Balança analítica BEL

- Conectar a balança à rede elétrica, ficando na condição de stand by (estado de espera), conforme mostra a Figura abaixo:

"Stand By" (estado de espera):

- Para levar a balança a condição de trabalho, pressionar a tecla **ON/OFF (L/D)**;
- Para retornar ao "Stand By", pressionar novamente a tecla **ON/OFF (L/D)**;

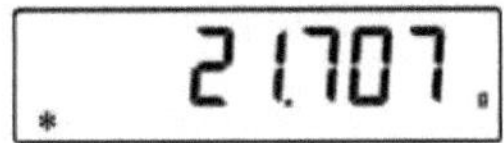

Pesagem simples:

- Depositar a amostra a ser pesada sobre o prato e ler o valor do peso no display quando o símbolo * (asterisco) de estabilidade aparecer;

Função Tara:

- Carregar um recipiente sobre o prato. No mostrador será exibido seu peso;

Revisão (Data): 06/01/2017	Elaborado por: Técnica de Laboratório Daiana Alves Machado	Página 1 de 2

- Pressionar a tecla **O/T**. Será exibida a palavra **"O-t"**;

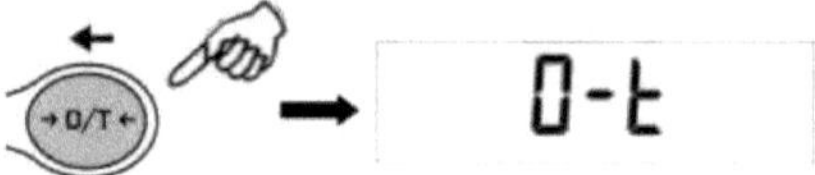

- Atingida a estabilidade será visualizado o valor **"0.000"**. No caso de não ser atingida a estabilidade por causa de correntes de ar, vibração ou outro tipo de distúrbio, a mensagem continuará a ser exibida;

- Introduzir os objetos no recipiente. Ler seu valor líquido no mostrador.

Limpeza:

- Antes de limpar a balança, desligue-a da tomada;
- Remover os restos de amostras e poeira com uma escova, pincel ou aspirador;
- Realizar a limpeza com pano úmido e detergente neutro. Não utilizar produtos de limpeza agressivos (solventes ou similares);
- Impedir a penetração de líquidos no aparelho durante a lavagem;
- Após a limpeza, passar um pano seco para retirar qualquer tipo de umidade.

yes
I want morebooks!

Buy your books fast and straightforward online - at one of world's fastest growing online book stores! Environmentally sound due to Print-on-Demand technologies.

Buy your books online at
www.morebooks.shop

Kaufen Sie Ihre Bücher schnell und unkompliziert online – auf einer der am schnellsten wachsenden Buchhandelsplattformen weltweit! Dank Print-On-Demand umwelt- und ressourcenschonend produziert.

Bücher schneller online kaufen
www.morebooks.shop

KS OmniScriptum Publishing
Brivibas gatve 197
LV-1039 Riga, Latvia
Telefax: +371 686 204 55

info@omniscriptum.com
www.omniscriptum.com

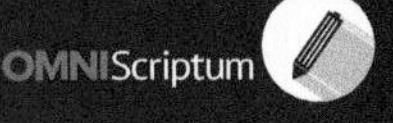